T0193166

Twin-Win Research:

Breakthrough Theories and Validated Solutions for Societal Benefit

Second Edition

Synthesis Lectures on Professionalism and Career Advancement for Scientists and Engineers

Synthesis Lectures on Professionalism and Career Advancement for Scientists and Engineers includes short publications that help students, young researchers, and faculty become successful in their research careers. Topics include those that help with career advancement, such as: writing grant proposals; presenting papers at conferences and in journals; social networking and giving better presentations; securing a research grant and contract; starting a company; and getting a Masters or Ph.D. degree. In addition, the series publishes lectures that help new researchers and administrators to do their jobs well, such as: how to teach and mentor, how to encourage gender diversity, and communication.

Twin-Win Research: Breakthrough Theories and Validated Solutions for Societal Benefit, Second Edition
Ben Shneiderman
September 2018

Oral Communication Excellence for Engineers and Scientists
Judith Shaul Norback
July 2013

A Practical Guide to Gender Diversity for Computer Science Faculty
Diana Franklin
April 2013

A Handbook for Analytical Writing: Keys to Strategic Thinking
William F. Winner
2013

University of Maryland (September 2017)

Twin-Win Research: Breakthrough Theories and Validated Solutions for Societal Benefit, Second Edition
Ben Shneiderman

ISBN: 978-3-031-01382-9 Paperback
ISBN: 978-3-031-02510-5 PDF
ISBN: 978-3-031-00327-1 Hardcover

DOI 10.1007/978-3-031-02510-5

A Publication in the Springer series
ON PROFESSIONALISM AND CAREER ADVANCEMENT FOR SCIENTISTS AND ENGINEERS # 4
Series Editors: Charles X Ling, University of Western Ontario and Qiang Yang, Hong Kong University of Science and Technology

Series ISSN 2329-5058 Print 2329-5066 Electronic

Twin-Win Research:

Breakthrough Theories and Validated Solutions for Societal Benefit

Second Edition

Ben Shneiderman
University of Maryland

SYNTHESIS LECTURES ON PROFESSIONALISM AND CAREER ADVANCEMENT FOR SCIENTISTS AND ENGINEERS #04

ABSTRACT

The thrill of discovery and the excitement of innovation mean that research is often immensely satisfying. But beyond the personal satisfaction, the goal of research is to improve the lives of people everywhere by driving revolutionary advances in healthcare, education, business, and government. This guidebook's strategies will help you shape your research and energize your campus so as to achieve the Twin Win: a breakthrough theory that's published and a validated solution that's ready for dissemination.

The action-oriented paths in this guidebook resemble a backpacker's guide to hiking. It suggests paths and gives you enough information to get started, while providing enough flexibility to take side treks and enough confidence to find your own way. Short-term projects include inviting speakers to campus, choosing appropriate research projects, and developing networking skills. Middle-term include seeking funding from government agencies and philanthropic foundations, sharpening your writing and speaking skills, and promoting teamwork in research groups. Long-term missions include changing tenure policies, expanding collaboration with business and civic partners, and encouraging programs that combine theory and practice.

KEYWORDS

research, Twin-Win Model, breakthrough theories, validated solutions, innovation, discovery

Contents

University of California at Irvine (March 2016)

Preface

This guidebook grew from years of discussions about an issue near to my heart: how to change the culture of college campuses in order to produce higher-impact research. Initially, those discussions led to a book, *The New ABCs of Research: Achieving Breakthrough Collaborations* (2016). Yet as I continued the conversation with colleagues and delivered more than 40 talks in the U.S., Canada, and the UK people kept asking me a question that I couldn't fully answer: "How can I change my own campus?"

So, I started preparing handouts. These one-pagers were meant for students, faculty, and academic leaders, like department chairs, institute directors, deans, and provosts. Memorable talks were at the Office of Science and Technology Policy and the National Science Foundation, where my one-page lists had a different set of items. This guidebook is a revised and combined version of those lists.

My focus is especially on North American universities and institutions that I know best, but I hope readers will find ways to translate these ideas into their national research ecosystems. My campus photos, which show the diverse universities I've visited, are meant to help readers appreciate that the ideas in the book can be widely applied.

The first edition of this book was a Kindle self-published book titled *Rock the Research: Achieving the Twin Win of Theory Breakthroughs and Societal Benefits*. Since my audiences have responded strongly to the Twin Win concept, I have made it the title for this second edition.

Cornell University (May 2003).

Acknowledgments

I'm grateful for supportive comments from, and lively discussions with, a range of colleagues. These people include Dan Sarewitz (Arizona State University), Lorne Whitehead (the University of British Columbia), Camille Crittenden (the University of California-Berkeley), Sameer Popat (the University of Maryland), Mark Western (the University of Queensland), and Mike Ash (the Better World Institute). I'm particularly grateful to my wife, Jennifer Preece, and to the many colleagues at the University of Maryland who offered valuable insights and encouragement. Hearty thanks also go to Asheq Rahman of Elsevier, who worked closely with me to provide the remarkable evidence from their SCOPUS database in Section 1.7. Finally, the meetings of the Highly Integrative Basic and Responsive Research Alliance helped push my ideas forward, as did help from the Government-University-Industry Research Roundtable of the National Academy of Sciences and the Association for Public and Land-grant Universities.

CHAPTER 1

Thinking about Research

1.1 INTRODUCTION

Research is great fun because innovation is exciting and discovery is a thrilling. But beyond the personal satisfaction, the goal of research is to improve the lives of people everywhere. Researchers have the opportunity to create astonishing innovations and make profound discoveries, which drive revolutionary advances in communication, healthcare, transportation, business, and government.

But to become this kind of superstar researcher, you have to be more than just lucky; you have to develop the right strategies. These strategies range from choosing problems to finding excellent collaborators, from validating your ideas to pushing back against skepticism. Learning and applying these strategies will increase the probability that your innovations and discoveries will blossom into commercial successes and influential theories that bear fruit as major societal benefits

Set your sights high! Your research could prevent cancer, cut energy consumption, or bolster cybersecurity. Your ideas could lead to highly cited papers, widely licensed patents, and successful business startups. It takes hard work, perseverance in the face of setbacks, polished social skills to push back against skepticism, and excellent presentation abilities to convey your success story.

But remember—dangers abound. Your ideas may fail, they may be bested by competitors, or they may be ignored because you failed to present them well. You also run the risk that the transformative changes you trigger may be disruptive for many people, may damage the environment, and may be appropriated by criminals, terrorists, and oppressive leaders. Research is a high-stakes endeavor, so the best researchers gird themselves for all possibilities.

Which is where this guidebook comes in. My goal is to help you redirect your research and to change your campus. If you can increase the impact of academic research, you and your colleagues can produce more potent innovations and more valuable discoveries.

The paths I outline can be pursued by individual students or faculty, or teams in a bottom-up fashion. These paths can also be important for top-down implementation by academic leaders like department chairs, deans, and administrators, as well as vice presidents of research, provosts, and even presidents. These leaders are typically the ones who promote visionary agendas described in ambitious strategic plans. And they often work closely with off-campus partners in business, government, nongovernmental organizations (NGOs), journalism, and beyond.

The bottom line: this guidebook is meant to resemble a backpacker's guide to hiking. It suggests paths and gives you enough information to get started, while providing enough flexibility to take side treks and enough confidence to find your own way.

1.2 THE TWIN WIN

The design of these paths is based on the ideas of many people, but it draws heavily from *The New ABCs of Research: Achieving Breakthrough Collaborations* (2016, http://www.cs.umd.edu/hcil/newabcs). For some people, discovery and innovation are separate pursuits. For a researcher, however, they're symbiotic. In fact, by pursuing them simultaneously, you're more likely to succeed at both.

And this parallel approach often leads to what I call the Twin Win. The Twin Win idea is to develop breakthrough theories in published papers AND validated solutions ready for widespread dissemination.

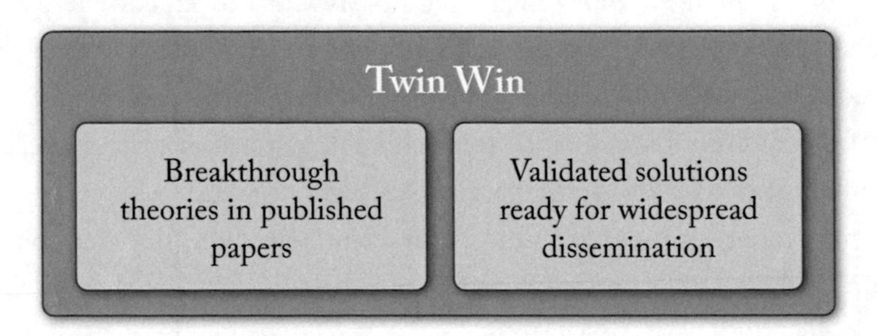

The idea of the Twin Win is so important that it's become the basis of a network of research leaders. The Highly Integrative Basic and Responsive (HIBAR) Research Alliance seeks to change campus cultures and has drawn support from the U.S. National Academies and the Association for Public and Land-grant Universities. (The latter hosts HIBAR's website: http://www.aplu.org/hibar.)

What kind of work does HIBAR do? Lorne Whitehead, of the University of British Columbia, offered this description: HIBAR specializes in projects that:

- seek both deep new knowledge and new practical solutions;
- use both academic research methods and practical design thinking;

- are led by both respected academics and real-world experts; and
- have long-term goals and short-term payoffs.

The HIBAR Research Alliance, with vice presidents of research and other campus leaders, held six meetings during 2017–2018 to discuss how to achieve the campus culture changes that are described in this guidebook.

Using the HIBAR approach to achieve the Twin Win isn't easy. That's because traditional academic attitudes and policies make it difficult to pursue fresh strategies. Old beliefs and established policies about how to do research are tough to shake, even in the face of growing evidence that new models are more reliable in producing large positive results.

Put another way: Changing the long-held convictions of your colleagues and mentors is a challenge—but it is possible. Changing institutional traditions and established policies is equally difficult, but these transformations are also possible. These changes are exactly what HIBAR Research Alliance seeks to engender.

The University of British Columbia, HIBAR Research Alliance (November 2017).

1.3 CHANGE

Most change-agent handbooks note that the first step of change is for those whose work will be changed to be aware of the big problems and to identify opportunities for improvement. For example, as a researcher, you need specific instructions about what steps to take. This clarity promotes willingness to give up familiar practices and try something new. Meanwhile, any improvements you make need to be measurable; this will reassure participants, even in the face of personal resistance and active opposition from others.

Additionally, because setbacks are inevitable, constant reassessment on your part is critical. Finally, recognition of your success from peers and superiors is essential to spread your ideas. Recognition also strengthens the commitment by others to new methods and goals.

If this all sounds hard, it *is*. As it should be. But remember: It's also possible.

Equally important: while these ideas are meant to help you achieve personal success and raise the impact of your lab or campus, they also have a larger impact. After all, each research project helps bring broader benefits to more people. Each small contribution is a tile in the mosaic of societal transformation.

1.4 CHECKLIST

If you want to become a visionary change agent in education, here's a checklist (derived from https://bit.ly/2MOyAO9).

1. **Alignment and Buy-in:** The change being considered should align with the overall values, vision, and mission of the initiative. Senior leadership must champion any new initiative. If someone at the C-suite level opposes the new initiative, it will likely die a slow and painful death.

2. **Advantage:** If the initiative doesn't provide a unique competitive advantage—preferably a game changing advantage—then it should at least bring you closer to an even playing field.

3. **Added Value:** Any new project should add value to existing initiatives. If it doesn't, it should show a significant return on investment to justify the dilutive effect of not keeping the main thing the main thing.

4. **Due Diligence:** Just because an idea sounds good doesn't mean it is. You should endeavor to validate proof of concept based upon detailed, credible research. Do your homework—put the change initiative through a

rigorous set of risk-reward and cost-benefit analyses. Forget this step and you won't be able to find a rock big enough to hide under.

5. **Ease of Use:** Whether the new initiative is intended for your organization, vendors, suppliers, partners, or customers, it must be simple and easy. Usability drives adoptability; therefore, it pays to keep things simple. Don't make the mistake of confusing *complexity* with *sophistication.*

6. **Risks:** Nothing is without risk, and when you think something is, that's when you're most likely to end up in trouble. All initiatives should include detailed risk-management provisions that contain sound contingency and exit planning.

7. **Measurement:** Any change initiative should be based upon solid business logic that drives corresponding financial engineering and modeling. Be careful of high-level, pie-in-the-sky projections. The change being adopted must be measurable. Deliverables, benchmarks, deadlines, and success metrics must be incorporated into the plan.

8. **The Project:** Many companies treat change as some ethereal form of management hocus pocus that will occur by osmosis. A change initiative must be treated as a project. It must be detailed and deliverable on a schedule. It must have a beginning, middle, and end.

9. **Accountability:** Any new initiative should contain accountability provisions. Every task should be assigned and managed according to a plan and in the light of day.

10. **Actionability:** A successful initiative cannot remain in a strategic planning state. It must be actionable through focused tactical implementation. If the change being contemplated is good enough to get through the other nine steps, then it's good enough to execute.

The University of British Columbia (September 2014).

1.5 AUDIENCE

This guidebook is geared toward campus participants. These people include the following groups:

- **students:** undergraduate and graduate;
- **faculty members:** assistant, associate, and full professors, as well as instructors, lecturers, adjuncts, post-doctoral researchers, research scientists, and related support staff;
- **academic leaders:** department chairs, center directors, deans, provosts, vice presidents of research, presidents, and others who shape academic life; and
- **administrators:** program directors, student advisors, development officers, public relations directors, physical plant managers, accounting staff, government relations specialists, and student service providers.

Working to change your campus's research could invigorate academic life, raise the reputation of your university, and increase the benefits for your city, state, or region, or even produce global benefits. Improving academic research is a reasonable goal because you can easily find partners among the well-defined community that is your campus. While research is the focus of this guidebook, changes to teaching, mentorship, and service will also be likely outcomes. In fact, since these components of campus life are richly interwoven, your progress on any path is likely to produce multiple advances.

Substantial and sustained changes will also require you to engage with off-campus business leaders and professionals, as well as government policymakers and agency staffers, all of whom hire students and fund research. Durable changes will also require you to engage with professional societies, journal and book publishers, conference organizers, and reporters. Finally, you'll want to seek out senior colleagues who can push for changes in government-funding agencies and philanthropic foundations, and business leaders. This guidebook describes collaborations that will help produce substantial and sustained campus changes.

Yale University (April 2011).

1.6 MESSAGING

I've spoken about these issues at more than 40 events, eliciting enough interest to get invited back and speak to other groups, but also generating pushback from two directions. About 10–20% of my audiences reject the premise that research should produce societal change. This cohort clings to the traditional belief that academic ivory towers are still the top place to work. They want to write theoretical papers and do laboratory studies, with little concern for impact and little interest in teaming up with business, government, and NGOs. I doubt this guidebook will change their minds.

Pushback also came from another 10–20% of my audiences who *support* these ideas. They felt that they were doing fine already in pursuing the Twin-Win goals of published papers and validated solutions. However, some of these sympathetic supporters still need to learn techniques for choosing problems, forming teams, and promoting their work. This guidebook could help these researchers find the right partners and adapt their strategies.

The remaining 60–80% consist of those who haven't thought much about these issues. For this cohort, I propose the following message.

> *"Evidence is growing that identifies new ways to improve the impact of your research. You might want to try it, but you are welcome to do what you have been doing. The strategy is to work on realistic problems and partner with professionals in business, government, and NGOs to bring your ideas to fruition."*

Under the Twin-Win strategy, you'll co-design a research plan, collaboratively implement it, and eventually co-author a paper. In other words, you're not only producing a breakthrough theory for publication; you're also producing a validated solution that's ready for dissemination. As a result, you stand to gain academic recognition *and* contribute mightily to society.

If you're interested in such a scenario, here are a few ways to proceed. Many faculty members have found that these partnerships produce continuing relationships, which advance research and disseminate it widely, while bringing substantial societal benefits. Often, the teams you assemble have uniquely powerful skills and access to potent resources that leads to remarkable discoveries and innovations. If you pursue the dual goals of producing breakthrough theories in publications and validated solutions that are ready for dissemination, you are more likely to gain academic recognition *and* make substantial societal benefits.

You could pursue this strategy by inviting relevant business, government, and NGOs researchers to speak on campus, encouraging your students to intern with them, and developing partnerships that lead to funding from them.

I've found that junior faculty and students are especially sympathetic to this message. Indeed, their biggest concern is where to start.

1.7 EVIDENCE

The evidence in support of the Twin-Win strategy is strong. My analyses began with searching for the impact of the University of Maryland research using the Elsevier SCOPUS database with 70M+ academic papers. I found that single author papers received, on average 3.0 citations, while co-authored papers with campus colleagues collected 6.1 citations. Co-authoring with colleagues in the U.S. produced 9.2 citations, while co-authoring with international partners raised the impact to 13.9 citations. However, the astonishing result was that papers with co-authors from businesses (and other large organizations) produced an average of 20.3 citations. While citations counts are flawed in many ways and they are not the only way to assess impact, this result validated the benefits of working with off-campus partners.

I wondered if this result was unique to my campus, so I began searching to see if the pattern held for other universities. A search for the top six private (Figure 1.1) and the top six public universities (Figure 1.2) showed remarkable consistency in the growing citation counts for papers with off-campus co-authors. The universities were chosen by their volume of publications in the SCOPUS database.

Further searching for universities in the United Kingdom (Figure 1.3) and Canada (Figure 1.4) provided confirmation of the thesis that corporate collaborations produced dramatically increased citation counts.

In interviewing academic colleagues who had many papers with corporate collaborators, I got enthusiastic comments about working with business, government, and NGO partners. A theory-oriented computer science colleague said "people in business have better problems," an attitude that was confirmed by a colleague working in computer vision. Having clearly focused problems to work on accelerated their work, while the resources and pressures of corporate efforts enabled these teams to produce meaningful results tied to realistic data. Validation or rejection for theories was quick and clear. Not all projects will have such positive outcomes, but there is an important lesson in this data: building connections with off-campus partners can lead to results that draw more attention from others.

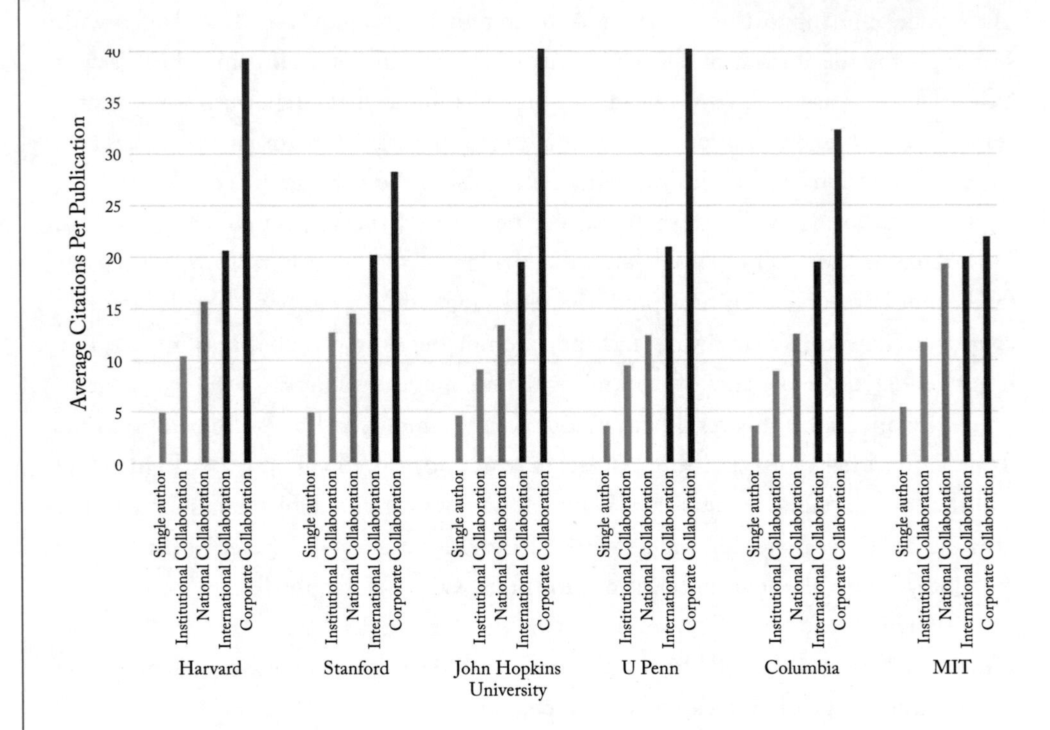

Figure 1.1: Effect of co-authorship on citations: top six private universities in the U.S.

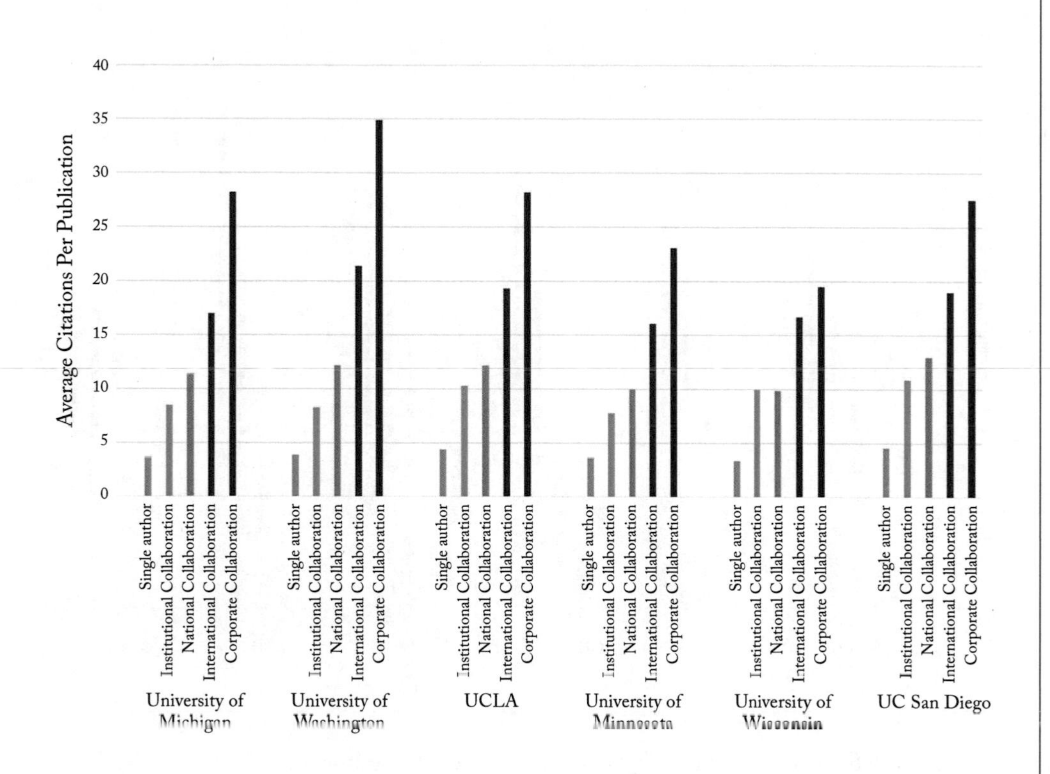

Figure 1.2: Effect of co-authorship on citations: top six public universities in the U.S.

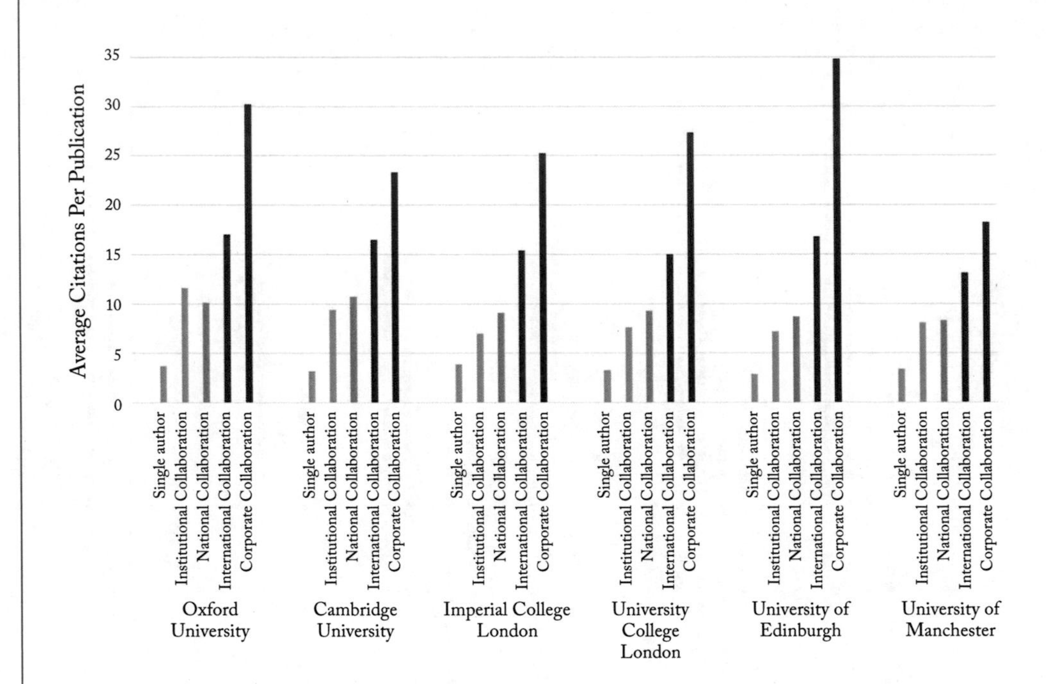

Figure 1.3: Effect of co-authorship on citations: six leading universities in the UK.

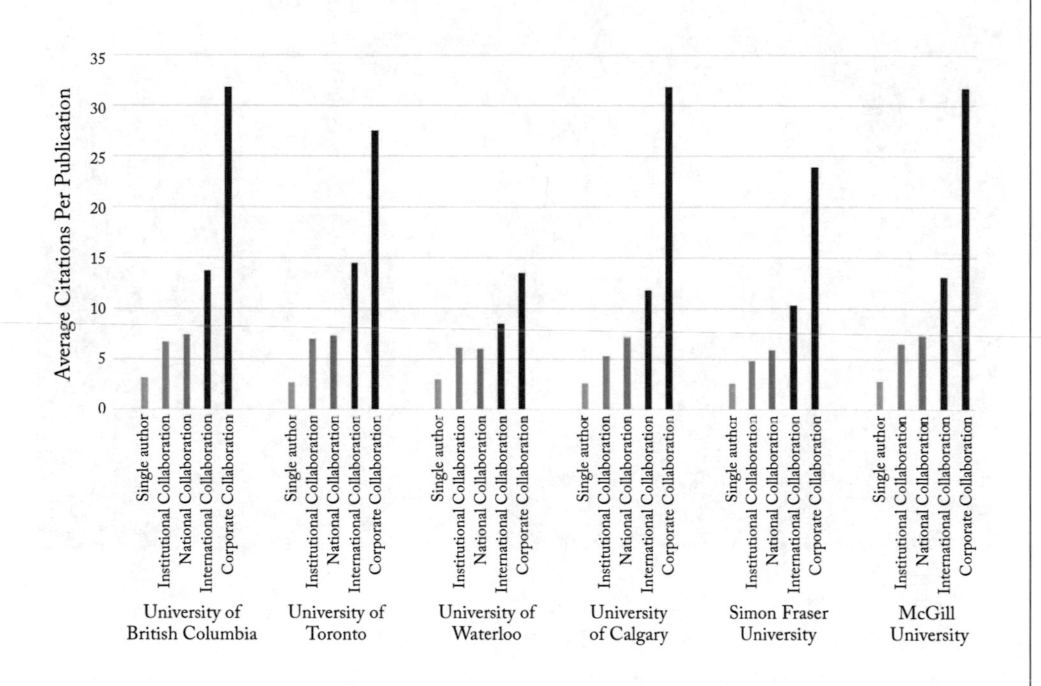

Figure 1.4: Effect of co-authorship on citations: six leading universities in Canada.

Cornell University (May 2003).

1.8 HISTORY

The idea of bringing academic researchers into closer contact with professionals who face authentic problems has long been discussed as a way to achieve higher societal benefits. The famed American poet and philosopher Ralph Waldo Emerson spoke in 1837 about academics uniting with farmers, businesspeople, and government. Emerson called for academics to engage in the real world: "Action … is essential … without it, thought can never ripen into truth." That encouragement remains valid today.

A century later, pressures to separate academic work from practical problems were generated by Vannevar Bush's manifesto on *Science: The Endless Frontier* (1945). Bush argued for a linear model, suggesting that basic research comes first, then applied research, and only then, considerations of commercial application.

The linear model was vigorously opposed by Tom Allen in the 1970s, Deborah Shapley and Rustum Roy in the 1980s, and many others afterward. An important contribution came from Donald Stokes' book, *Pasteur's Quadrant: Basic Science and*

Technological Innovation (1997), which proposed a fresh strategy: "use-inspired basic research." His reference to Pasteur reminded readers about Pasteur's work on the problems of vintners and dairy farmers, which produced the Twin Win of solutions to their problems and the germ theory of disease. Lewis Branscomb's (2007) essay supported the idea that creativity and utility (basic and applied) research were happy partners.

In recent years, my book, *The New ABCs of Research: Achieving Breakthrough Collaborations* (2016), advocated for "applied and basic combined." I was joined by Narayanamurti and Odumosu, who wrote *Cycles of Invention and Discovery: Rethinking the Endless Frontier* (2016). In the same year, Dan Sarewitz published a powerful essay on "saving science," which advocated that scientists increase their impact, while reducing the prevalence of results that can't be replicated. Sarewitz stressed that "scientists must come out of the lab and into the real world." A similar call for emphasizing applications as the path to discoveries came from a group of information-visualization researchers, who called on their colleagues to "apply or die" (Weber et al., 2017).

Leading organizations have also sounded the reform alarm. They've invoked many terms to describe their variations on the theme of blending applied and basic research through interdisciplinary, cross-disciplinary, multidisciplinary, and transdisciplinary approaches (ARISE2, 2013). For example, the National Research Council has called for *convergence* (2014), while the National Academy of Engineering (2017) emphasizes "work[ing] in concert to maximize the value created for society." Meanwhile, the Irish Research Council (2016) uses the phrase "engaged research" to describe collaborations "*with* community partners, rather than *for* them."

A word of caution: I do not think that interdisciplinary or convergent ideas are sufficient to achieve the goal of high research impact. Such approaches may be helpful, but a vital component is that researchers need to work with professionals who have authentic problems, which allow for validation of proposed theories and solutions. Working on authentic problems is also the theme of the growing community of supporters of "implementation science."

The importance of developing the strategies for team formation and management was presented very effectively in a thoughtful, well-documented, and highly readable report on *Enhancing the Effectiveness of Team Science* (Cooke and Hilton, 2015).

Michel M. Crow, President of Arizona State University, has worked for more than a decade to steer his campus to transform society by conducting "use-inspired research" that is "socially embedded" in local, regional, and national projects (Crow and Dabars, 2015). Shirley Ann Jackson (2014), President of Rensselaer Polytechnic Institute called for her faculty to "collaborate more effectively with businesses and governments… educating our students in multidisciplinary and collaborative thinking… guided by social concerns and ethics."

All these reports are helpful in understanding the broad push to change academic research culture, but there are those who resist and seek to preserve current practices. This guidebook offers a set of ways for you to make successful sustained changes. You may want to keep up with a group of academic leaders who have been meeting to develop strategies that can work across many campuses and disciplines. This determined set of leaders has adopted the term Highly Integrative Basic and Responsive (HIBAR) Research Alliance. Among the alliance's accomplishments, it has established relationships with prestigious organizations such as the Association for Public and Land-grant Universities and the National Academies of Science, Engineering, and Medicine. You can learn more about HIBAR here: http://aplu.org/hibar.

The University of British Columbia, HIBAR Research Alliance (October 2017).

1.9 FIRST STEPS

The first step in making a change is to become aware of the new possibilities. For researchers who have always worked on theoretical problems and in laboratories, the idea of working with practitioners in the field can be troubling. However, many professionals working for business, government, or NGOs are well-trained experts: they sometimes have doctoral degrees and usually have a deep understanding of interesting problems. What's more, these people are delighted to meet academics who want to explore real-world problems. Of course, academic goals are often different from business goals, so discussion is needed to identify problems that have potential for the Twin Win.

Some academics fear that business professionals want free consulting, help with well-understood problems, or academic seals of approval for questionable business agendas. Other issues to resolve include project goals, schedules, commitment of effort, and outcomes. Discussing shared intellectual-property ownership and publication plans at the beginning will help ensure happy endings. Publications are vital to academic success, so agreements that ensure faculty right to publish are essential. In our lab, we typically give our partners 10–14 days to review papers to check the contents and make suggested changes, but we are free to go ahead to submit the research. The desire for continuing the partnership is the most important ingredient in reaching satisfying agreements.

When businesses wanted full ownership of the outcomes and veto power in publication, we simply walked away from that partnership, even when they offer to provide financial support. Most businesses have come to understand the way universities operate, so they are easier to work with now, than a few decades ago. Often businesses are so eager to work with academic partners that they will support the research, provide staff time, eagerly hire students, and offer consulting contracts for faculty members.

Working with local, state, or national governments, or NGOs, is another fruitful path to high-impact research. Their motivations differ from that of businesspeople, but they are often eager to draw on expertise found in universities. I was especially proud of a group of students in my graduate course who worked to improve the data presentation for the U.S. Bureau of Labor Statistics (BLS), which produces the monthly jobs and unemployment reports. At the end of the semester, the students travelled to Washington, DC to present their paper and prototype to the Commissioner and her staff. As a result, a few months later, BLS changed its website.

The main early step for faculty members is to define a problem with the potential for a Twin Win. Having clear goals is important, but there are two related concerns: you must be flexible in refining, or even redefining, your goals, and you must plan your work to allow for publication can in three to nine months. Early-stage publications could define the problem, build a taxonomy of issues, curate a dataset to support research, and produce a thorough literature review that brings clarity to an emerging topic. Middle-stage outcomes could be new software tools, improved chemical processes, validated social surveys, advanced metrics, or enhanced mathematical models.

These contributions can be applied to known problems to produce actionable results that lead to further publications. At the same time, these contributions can be disseminated to colleagues for them to apply, validate, and improve your methods. In later stages, refined methods are often applied to new problems, bringing even wider impact and interest. If the partnerships go well, then convening a workshop at a conference, writing a summary survey article, or licensing your novel methods to others becomes possible.

Some of the paths presented in this guidebook will help junior and senior researchers to change their ways of working. Motivated researchers may also be eager to engage with colleagues to alter the ways of working in their lab, department, or campus.

Stony Brook University (May 2015).

1.10 BOTTOM-UP AND TOP-DOWN APPROACHES

Achieving substantial and sustained campus change is best done by combining bottom-up and top-down approaches. A good way to start is bottom-up—by convening discussion groups among 5–20 faculty and students who believe in the Twin-Win agenda (of breakthrough theories in published papers *and* validated solutions that are ready for dissemination).

Lorne Whitehead, of the University of British Columbia, is a master of faculty discussions. His breakout groups give everyone a chance to speak, while eliciting diverse experiences. For example, he gathered 120 faculty members over a year as he sought to understand what worked on his campus in areas such as forest management, affordable medical diagnostics, and urban resilience. Whitehead's evolving manifesto provided background theory, local stories, and a reading list for newcomers. I was impressed when I read his manifesto, and fascinated when I attended two of his meetings.

Inspired by what Lorne had accomplished, my colleague Rob Briber and I held a roundtable meeting of 15 University of Maryland (UMD) faculty members. The diverse group from across UMD colleges produced the most interesting discussion in all my years on campus. These highly successful researchers consistently described their strategies of working with off-campus partners to take on real problems, like ensuring food safety, improving wi-fi effectiveness, and fighting terrorism.

Another way of reaching out to colleagues and students around campus is to give talks to small groups of 10–25 people to describe the Twin-Win concept. As I was testing out the ideas in my book, I used the term "applied and basic combined," which is one of the meanings of *The New ABCs of Research*. Other ABC terms that I described were "achieving breakthrough collaborations" for forming teams and "always build connections" for networking to spread the news about your early research.

Supportive responses from my audiences led to further invitations. By now, I've given more than 20 talks just at the University of Maryland, in addition to more than 20 talks at campuses and government agencies in the U.S., Canada, and the UK.

Other bottom-up approaches are to arrange seminar groups, circulate papers, hold workshops, and deliver tutorials. These approaches build communities with a common ground of shared terminology and mutually held beliefs, produce new connections, and promote fresh thinking. And, best of all, they involve minimal effort and low risk. You will have to decide which paths are easy for you to accomplish and most suited for your campus.

Working bottom-up helps build interest and connect colleagues, but at some point, a top-down approach is probably necessary to produce broad changes across campus. The top-down approach requires the participation of campus leaders who can weave these ideas into strategic plans, gain support from campus leaders, and promote debates at a Campus Senate or similar body. Top-down approaches are needed to change hiring agendas, revise tenure and promotion policies, and allocate resources to researchers who form Twin-Win collaborations.

The idea of a Faculty Research Communicator Award (https://research.umd.edu/communicatoraward) drew a warm reception from the Vice President for Research at the University of Maryland. The award was designed "to recognize researchers who take a proactive approach to sharing their research or discussing research directions with the public." In just a few weeks, the announcement was prepared, then nominations were received, and, in a few more weeks, the award ceremony honored four faculty members (https://research.umd.edu/news/news_story.php?id=9294).

Making this award an annual event increased awareness that faculty are expected to do more than merely publish research papers. Telling their story in ways that welcome a wider circle of people raises appreciation of their contributions, which could increase interest and financial support.

As you draw more attention from bottom-up activities, you will also gain credibility to solicit top-down efforts from campus leaders. You can invite department chairs and deans to adopt these ideas as part of their own plans. They may have ideas to increase teamwork, introduce courses about design thinking, or promote entrepreneurship. On my campus, leaders actively encourage commercial applications by way of the National Science Foundation's I-Corps (Innovation Corps); business start-up efforts through the Colleges of Agriculture, Business, and Engineering; and partnerships with businesses, government agencies, and NGOs. Well-funded organizations on campus and in the State of Maryland, as in most states, support academic partnerships with businesses and commercialization of research outcomes.

Academic leaders can also be influential by presenting an inspirational vision, repeating it consistently, and tying each success to the larger goals of their campus. These leaders can change hiring, promotion, and tenure policies to encourage and reward teamwork across disciplines, collaborations with practitioners, and successes that go beyond publishing papers (e.g., patents, start-ups, policy changes, education, etc.).

The University of British Columbia (September 2014).

1.11 SHORT-, MEDIUM-, AND LONG-TERM STRATEGIES

Whether you're interested in exploring ways to change your research, or ready to take on the larger challenge of changing your campus culture, it's important to begin by learning what's happening on your own campus. Your Vice President of Research, or similarly titled leaders, is probably already trying to raise the impact of campus research. Their ideas, resources, and funding could help you get started, and you could help them promote campus-culture changes.

Most universities also have units that promote entrepreneurial activities, build bridges to local businesses, and help guide researchers to produce patents, copyrights, and other intellectual property. For example, my campus has an active Office of Technology Commercialization that provides "expert guidance, support, and assistance in safeguarding intellectual property, encouraging research, facilitating technological transfer, and promoting collaborative research and development agreements with industrial sponsors."

You may also want to learn about national organizations that promote partnerships for high-impact research (see Section 1.12). A favorite of mine is the University-Industry Demonstration Partnerships (UIDP). UIDP describes itself as "an organization where representatives from some of the finest innovation companies and best research universities in the world meet and commit to active participation in pursuit of excellence in university-industry collaboration and partnership. We share a commitment to practical problem-solving and a belief that together we can overcome challenges we could not alone." UIDP runs events that bring researchers and businesspeople together and provides templates for agreements that facilitate collaboration.

The University of North Carolina at Chapel Hill (September 2016).

1.12 ORGANIZATIONS THAT PROMOTE HIBAR RESEARCH AND TWIN-WIN STRATEGIES

Here's a list of organizations you should know about, along with links to their websites:

- **Association of Public and Land-grant Universities:** http://www.aplu.org/hibar;
- **Business-Higher Education Forum:** http://www.bhef.com/about;

- **Coalition for Networked Information:** https://www.cni.org;
- **Council for Financial Aid to Education:** http://cae.org/about/history/;
- **EDUCAUSE:** https://www.educause.edu;
- **Government-University-Industry Research Roundtable:** http://sites.nationalacademies.org/PGA/guirr/PGA_081653;
- **National Alliance for Broader Impacts:** https://broaderimpacts.net/about/;
- **National Organization of Research Development Professionals:** http://www.nordp.org/conferences;
- **State Science and Technology Institute:** http://ssti.org/aboutSSTI; and
- **University-Industry Demonstration Partnership:** https://www.uidp.org.

The University of California at Los Angeles (February 2009).

1.13 REFERENCES

Here's a brief history of the growing movement of academics who seek the Twin Win in their research:

1. Allen, T. J. (1977). *Managing the Flow of Technology: Technology Transfer and the Dissemination of Technological Information within the R & D Organization*. MIT Press, Cambridge, MA.

2. American Academy of Arts and Sciences. (2013). *ARISE2: Advancing Research in Science and Engineering, Unleashing America's Research & Innovation Enterprise*. Cambridge, MA. 15

3. Branscomb, L. (2007). The false dichotomy: Scientific creativity and utility. *Issues in Science and Technology* 16, 1. http://www.issues.org/16.1/branscomb.htm. 15

4. Bush, V. (1945). Science: The Endless Frontier. A report to the President on a program for postwar scientific research, Office of Scientific Research and Development, Washington, DC. DOI: 10.21236/ADA361303. 14

5. Cooke, N. J. and Hilton, M. L. (Editors) (2015). *Enhancing the Effectiveness of Team Science*. National Academies Press, Washington, DC. http://www.nap.edu/catalog/19007/enhancing-the-effectiveness-of-team-science. 15

6. Crow, M. M. and Dabars, W. B. (2015). *Designing the New American University*. Johns Hopkins University Press. 15

7. Goldstein, A. P. and Narayanamurti, V. (2018). Simultaneous pursuit of discovery and invention in the US Department of Energy. *Research Policy* 47, 8, 1505–1515. DOI: 10.1016/j.respol.2018.05.005.

8. Irish Research Council. (2016). *Engaged Research: Society & Higher Education Addressing Grand Societal Challenges Together*. Dublin, Ireland. 15

9. Jackson, S. A. (2014). Op-ed: The new polytechnic: Preparing to lead in the digital economy. *US News and World Report* (September 22, 2014). https://bit.ly/Z9sSKv. 15

10. Narayanamurti, V. and Odumosu, T. (2016). *Cycles of Invention and Discovery: Rethinking the Endless Frontier*. Harvard University Press, Cambridge, MA. DOI: 10.4159/9780674974135. 15

11. National Research Council. (2014). *Convergence: Facilitating Transdisciplinary Integration of Life Sciences, Physical Sciences, Engineering, and Beyond*. The National Academies Press, Washington, DC. 15

12. National Academy of Engineering, Committee on a Vision for the Future of Center-Based Multidisciplinary Engineering Research. (2017). *A New Vision for Center-based Engineering Research*. Washington, DC: The National Academies Press, Washington, DC.

13. Sarewitz, D. (2016). Saving science. *The New Atlantis: A Journal of Technology & Society*. 49, 4–40.

14. Shapley, D. and Roy, R. (1985). *Lost at the Frontier: U.S. Science and Technology Policy Adrift*. ISI Press, Philadelphia, PA.

15. Shneiderman, B. (2016). *The New ABCs of Research: Achieving Breakthrough Collaborations*. Oxford University Press. DOI: 10.1093/acprof:oso/9780198758839.001.0001. 15

16. Shneiderman, B. (2018). Twin-Win Model: A human-centered model for research success. *Proceedings of the National Academy of Sciences*. To appear October 2018.

17. Spector, A., Norvig, P., and Petrov, S. (2012). Google's hybrid approach to research. *Communications of the ACM* 55, 7, 34–37. DOI: 10.1145/2209249.2209262.

18. Stokes, D. (1997). *Pasteur's Quadrant: Basic Science and Technological Innovation*. Brookings Institution, Washington, DC. 15

19. Weber, G. H., Carpendale, S., Ebert, D., Fisher, B., Hagen, H., Shneiderman, B., and Ynnerman, A. (2017). Apply or die: On the role and assessment of application papers in visualization. *IEEE Computer Graphics & Applications* 37, 3. DOI: 10.1109/MCG.2017.51. 15

The University of Maryland (September 2017).

CHAPTER 2

A Starter Set of Paths to Enhance the Quality and Impact of Your Research

So, how do you get started in changing your campus? The second part of this guidebook outlines 34 paths for students, faculty, academic leaders, and administrators to enhance the quality and impact of their research. (You can think of these paths as *short walks*, *weekend hikes*, and *month-long treks*, although I've labeled them formally as "projects," "efforts," and "missions.") Where appropriate, I've included links for further information. The goals are to pursue Twin-Win strategies to increase the impact of research on society.

2.1 SMALL, SHORT-TERM PROJECTS

"Something small soon" is my favorite phrase in guiding students to make progress in their projects. Small goals can be achieved relatively quickly, thus building a student's confidence, helping him/her engage with others, and equipping him/her with a better sense of what to do next

The University of Washington (October 2016).

2.1.1 INVITE A SPEAKER

A simple first step in learning about a topic, building a collaboration, and stimulating interest is to invite a speaker from your campus or beyond to give a talk about his or her research. Once the date and time are agreed on, you'll have to reserve a room and publicize the event.

But your role as the organizer is more than getting a speaker and an audience. To understand the range of possibilities, start by making a list of potential speakers. Consider the impact their topic will have on the audience you want to attract. Then, work with the speaker to compose a clear, compelling title, accompanied by an informative abstract; this will help create buzz.

Your announcement can go out one to three weeks in advance, mostly to targeted email lists and via Facebook posts; personal invitations to key individuals and groups add to the interest. Sending links to key papers or the speaker's homepage enable potential attendees to learn more.

When you introduce the speaker, be sure to indicate why you have chosen this person and topic for your audience, and describe the outcomes you hope for. Indicate what aspects of the speaker's work are relevant to your audience, then suggest what collaborations might emerge. At the end, invite questions, but be prepared to ask some on your own.

Follow up a day later by thanking your speaker and by encouraging attendees to pursue the outcomes you hoped to inspire. Your metrics of success can range from the number of attendees, to the number of questions asked, to the adoption of research methods or software tools. Sometimes, however, the formation of a single collaboration is a sufficiently powerful result that justifies the entire effort.

2.1.2 RUN A READING GROUP

Invite faculty and students to meet once a week or once a month to read and discuss papers on a research topic of interest. The topic could be an established problem in which there is fresh progress, or something emerging and hot. Generate a discussion to choose the paper and moderator for each meeting. Announce the meetings by email, Facebook, or website, and invite relevant experts from on-campus and off.

An effective strategy is to have a leader for each reading group introduce the paper, invite discussion, and ask provocative questions. Try to have every attendee make at least one comment before anyone makes a second comment. Move the discussion forward from comments on the paper to suggestions of its implications for your attendees.

For example: How might it change research? What did the authors fail to do? What new directions does their work open up?

Afterward, your group might write a short summary of the discussion, along with a few interesting questions for the author of the paper, so as to start a dialog. This might lead to an invitation to visit campus. Sending the summary to those who could not attend keeps your community moving forward in a coordinated way.

2.1.3 ORGANIZE A WORKSHOP

If there's interest in a topic or problem, then having a half-day or full-day workshop will ignite that spark, which might lead to formation of collaborations. You can organize a workshop with 3–10 speakers, each of whom will describe their current research on the topic at hand, so that participants can learn about different approaches, discuss open problems, and share information on potential research funders. Your workshop can produce a shared bibliography, list of current researchers, and plans for a follow-on event that might bring together even more researchers. Workshops can be fun if you plan to social aspects carefully, such as informal meals, breakout groups to work on problems, and lists of email contacts to promote future connections.

Once you've gotten energized by doing a day-long workshop, you could work with colleagues to run longer workshops and summer courses. Matt Salganik reviews his two-week workshops held at several universities with a thoughtful post-mortem analysis.

University of Maryland Human-Computer Interaction Lab (September 2016).

2.1.4 OFFER A TUTORIAL TO BUILD RESEARCH SKILLS

Many research projects require special skills, such as use of Geographic Information Systems (GIS), statistical methods, laboratory procedures, computer programming, data curation, survey design, and much more. Sometimes, broadening your thinking about research methods can open up possibilities; qualitative, quantitative, mixed, ethnographic, or case-study methods often enrich projects. The good news is that these skills are valuable in many careers, will likely increase the students' employability, and may spawn communities with shared interests. You could invite an expert to offer a half-day, weekly, or weeklong tutorial.

2.1.5 ORGANIZE FACULTY MEETINGS TO DISCUSS TWIN-WIN RESEARCH STRATEGIES

A major bottom-up strategy for campus culture change is to bring together successful faculty researchers to discuss their strategies. Lorne Whitehead, of the University of British Columbia, inspired me with his success, leading me to invite 15 of my colleagues for a two-hour discussion in May 2017. To my amazement, all 15 came, leading to a spirited and fascinating discussion. (I'm including below an edited version of the invitation I sent.) Smaller-group discussions over lunch or other informal occasions could be helpful in promoting awareness and interest in Twin-Win research strategies.

Here's an email I sent to 15 colleagues inviting them to a discussion about increasing the impact of research.

> Dear Colleagues,
>
> I'd like to invite you to a discussion about increasing the impact of research. Here are the details:
>
> **Goal**
>
> Develop common strategies across campus units that promote Twin-Win partnerships for breakthrough research and societal benefits.
>
> Many campuses have launched programs of "engaged research," "innovation and entrepreneurship," and "service to society." In fact, there are several relevant efforts at the University of Maryland. This first meeting is meant to launch a bottom-up effort to share experiences among faculty so as to coordinate efforts. Our meeting will provide input for the Associate Vice President for Innovation and Economic Development, who will be leading an initiative to make a comprehensive plan.
>
> **Issues**
>
> 1. Partnerships: Develop strategies to facilitate Twin-Win partnerships to produce breakthrough research and economic development. What changes are needed to facilitate partnerships between campus researchers and business leaders (national and local); government partners (cities, counties, state, federal); and nongovernmental organizations? Which template agreements work? How much seed funding is needed? Who does the outreach?

2. Policies. Revise campus hiring, tenure, and promotion policies to support teamwork that includes off-campus partnerships and on-campus collaborations.

    - Document collaboration and team leadership for tenure cases.
    - Collect impact measures for research publications by citation counts, downloads, media mentions, etc.
    - Report evidence that includes research publications, media visibility, internal and external presentations, industry uptake, undergraduate and graduate educational impact, leadership in professional societies, influence on national, state and local policies, etc.
    - Encourage faculty visibility by way of blog posts, Wikipedia articles, social media, traditional media, internal and external presentations, op-eds, etc.

3. Processes. What steps are most needed at the University of Maryland to promote Twin-Win partnerships?

    - Facilitate meetings—become a convener for the community. Hire staff person?
    - Generate email lists to inform outsiders about campus events.
    - Improve campus social media presence.
    - Improve campus facilities for meetings, simplify transportation, provide food.
    - Recognize faculty who produce results.

2.1.6 STUDY MEANINGFUL PROLEMS

Choose major problems, and then find a smaller piece for which your skills enable you to make a substantive advance. Realize that many grand challenges have sociotechnical components, requiring new forms of research, data collection, and hypotheses testing. Here's a list of problems and sponsors:

1. **National Academies Grand Challenges:** http://www.engineeringchallenges.org;
2. **United National Sustainable Development Goals:** https://sustainabledevelopment.un.org;
3. **Bill and Melinda Gates Foundation, Grand Challenges:** http://gcgh.grandchallenges.org;
4. **National Science Foundation, Computer and Information Science and Engineering:** https://nsf.gov/dir/index.jsp?org=cise; and
5. **National Science Foundation, Engineering:** http://www.nsf.gov/dir/index.jsp?org=ENG.

The University of California at Irvine (February 2009).

2.1.7 FORM A RESEARCH PROJECT

When a group of faculty and students recognizes the importance of a research problem, it can form a project to work on the problem. Initially the work may be voluntary and mainly devoted to learning about the problem, previous work, and possible new approaches. As the project team begins to team up, then funding may be needed to enable faculty and students to spend ample time on the project and to acquire the equipment or resources needed. Funding may also be needed to attend conferences and to present results. Yet even without funding, you can often get a lot done to demonstrate the value of your ideas, build the team, and start publishing papers. Many of my influential innovations were done in collaboration with undergraduate and graduate students working on class assignments or independent study projects.

2.1.8 DEVELOP YOUR SOCIAL MEDIA FLUENCY

You may already know about social media such as Facebook, Twitter, YouTube, Wikipedia, and blogs. But learning to use them to advance your research may take some reading and learning. Take a look at these (http://www.sciencemag.org/news/2014/09/top-50-science-stars-twitter) AAAS lists (http://www.sciencemag.org/news/2014/10/twitters-science-stars-sequel) of science stars on Twitter, as well as this follow-up analysis (https://bit.ly/2htP5xU). And then check out this list of top 100 science blogs (https://blog.feedspot.com/science_blogs/).

No one can be active on every social media platform, so pick a few and spend one hour a month cultivating your online presence. That amount of time will gain you greater visibility and help you think more clearly about what your impact is and how to increase it.

The following guides to promoting your research will offend some traditionalists, but they contain valuable messages for researchers in shaping their story and reaching their audiences:

1. **The Becker Model:** Strategies for Enhancing the Impact of Your Research: https://becker.wustl.edu/impact-assessment/strategies;

2. **10 Simple Steps for Building a Reputation As a Researcher in Your Early Career:** https://bit.ly/2Neq4Vz;

3. **How to Write a Blog Post from Your Article in 11 Easy Steps:** https://bit.ly/1VkWr2b; and

4. **Marketing for Scientists: How to Shine in Tough Times**, by Mark J. Kuchner.

Iowa State University (April 2005).

2.1.9 DEVELOP YOUR NETWORKING SKILLS

Build relationships with leaders in your field, so they can learn about your work and you can possibly collaborate with them. Invite them to speak at your campus. Send a copy of your draft paper to five or more leaders in your field to ask them if you've been fair in citing their work and if they've done any more recent work that you should know of. Conduct a weekly reading group on campus for those who share your interest. You will learn about related work and meet potential collaborators. Run a lecture series or a 20–40-person workshop for researchers working in a specific area of your field. Become involved as a member of a program committee or editorial board.

1. **5 Steps to Seriously Improve Your Networking Skills:** https://www.entrepreneur.com/article/245995.

No matter what profession you're in, networking is the fuel that accelerates success. Not only is it useful for learning directly from individuals you meet, but the benefits of association and growing your own authority are just as powerful.

2. **10 Ways to Improve Your Networking Skills:** https://bit.ly/2NgfRIf.
3. **The Art of Networking:** https://uk.hudson.com/networking-skills.

2.2 MEDIUM-SIZED, MIDDLE-TERM EFFORTS

Doing "something small soon" will help you make progress, win over collaborators, and clarify what needs to be done next. Here are a few more substantial efforts to help guide you.

Simon Fraser University (September 2016).

2.2.1 ARRANGE A SPEAKER SERIES

Bring off-campus speakers from business, government, and/or NGOs to describe their research, and offer 2–4 research problems that could lead to an academic collaboration. These talks will inform students and faculty about current problems and may lead to funding, student internships, or employment for grad students. An invitation to speak on your campus will be seen as an honor and an opportunity, but make sure your invited speaker has a research orientation and is an experienced speaker. Choose an appropriate room for your guest; this can be a conference room with 10–12 seats or a larger classroom with 20–40 seats. Use email lists to make announcements to the right set of students and colleagues. Once you gain experience, you can use larger meeting rooms or auditoriums.

The University of British Columbia (March 2016).

2.2.2 MAKE IT EASY FOR OFF-CAMPUS GUESTS TO VISIT

Make it easy for guests to visit your campus. Include information on public or campus bus or rail routes in your announcements and websites. Promote the use of bicycles by putting a map with biking routes on your website and making sure there are bicycle racks near the event. Give clear information about how to get to the event by car or foot. Over time, you can work with regional transportation organizations and on-campus services to develop more frequent and high-quality services, as well as public parking for private cars.

2.2.3 FACILITATE MEETINGS ON CAMPUS

Make it easy for faculty, departments, and colleges to hold public events on campus. Similarly, make it easy for organizers of these events to reserve rooms online or with a central phone contact. Develop simple strategies for organizers to provide catered food, audiovisual services, and the right furniture. When your campus becomes a convener for local professionals, visitors from other campuses, and the public, more ideas can flow across networks. And the goodwill from being the convener could translate into more donations for your campus, or even more support from city or state agencies.

Meeting at the University of Maryland (May 2018).

2.2.4 SEEK FUNDING FROM GOVERNMENT AGENCIES

Most university research in the U.S. is funded by government agencies through grants devoted to specific topics. Proposals to the National Science Foundation (https://www.nsf.gov), the National Institutes of Health (https://www.nih.gov), and other agencies are very competitive, with success rates usually under 20%, and sometimes lower. This means that many researchers get used to the proposal process and submit multiple grants. The peer review process takes six to nine months, and sometimes produces disturbing comments from reviewers, but don't be put off. Reviewers may make mistaken assumptions about your proposal, but consider each negative comment as valuable feedback in learning how to make your submission better. Grants from these prestigious agencies get high respect from colleagues, but grants from other agencies are also valuable, often bringing researchers close to real problems and developing productive partnerships.

You can find more information from the American Association for the Advancement of Science (http://www.sciencemag.org/careers/where-search-funding), as well as the federal government (https://www.grants.gov). And here are some sobering thoughts about getting grants (https://bit.ly/2MOLcoh).

Massachusetts Institute of Technology Media Lab (May 2010).

2.2.5 SEEK FUNDING FROM PHILANTHROPIC FOUNDATIONS

An increasing amount of research is supported by private philanthropies, which are often established by an individual or family. There are many of them to explore, but your university research office is likely to have a list of foundations that they have worked with. Here are a few of the well-known ones:

1. **Carnegie Foundation:** https://www.carnegie.org;
2. **Ford Foundation:** https://www.fordfoundation.org;
3. **Guggenheim Foundation:** https://www.gf.org;
4. **MacArthur Foundation:** https://www.macfound.org;
5. **Simons Foundation:** https://www.simonsfoundation.org; and
6. **Sloan Foundation:** https://sloan.org.

Don't overlook other foundations that are local or especially interested in your kind of research. Check with your campus development office for contacts with foundations that are already working to support your campus. For information on getting started with grants, visit http://grantspace.org/tools/knowledge-base.

Working with foundation program managers can be satisfying, as they tend to take a strong interest in the research, even making constructive suggestions and putting you in touch with related projects.

The University of North Carolina at Chapel Hill (January 2005).

2.2.6 SHARPEN YOUR WRITING SKILLS

Refine your writing skills to tell your research story more persuasively. Don't just describe what you've done; make your story relevant to others by explaining how they can benefit from your work. Clarify what your contributions are. Give names to central ideas, draw figures to illustrate processes, show data with lucid charts, and produce compelling photos of your work. In addition to your refereed research publication, tell your story on webpages, in blogs, with videos, on Wikipedia, and in tweets.

1. GreatResearch: Story-Telling 101: Writings Tips for Academics. http://greatresearch.org/2013/10/11/storytelling-101-writing-tips-for-academics/.

2. Dupre, L. (1998). *Bugs in Writing: A Guide to Debugging Your Prose*, Second Edition. Addison-Wesley.

3. Hartley, J. (2008). *Academic Writing and Publishing: A Practical Handbook*. Routledge.

4. Jones, S. P. *How to Write a Great Research Paper*. YouTube video, 34 minutes. https://youtube.com/watch?v=g3dkRsTqdDA.

5. Lange, P. How to write a scientific paper for peer-reviewed journals (undated). http://www.ease.org.uk/publications/ease-toolkit-authors/how-to-write-a-scientific-paper-for-peer-reviewed-journals/.

6. Luey, B. (2009). *Handbook for Academic Authors*, Fifth Edition. Cambridge University Press. DOI: 10.1017/CBO9780511807893.

7. Ramsey, N. (August 2014). *Teaching Technical Writing Using the Engineering Method: A Handbook for Groups*. Tufts University. CS-Oriented Guide, Free PDF. http://www.cs.tufts.edu/~nr/pubs/eng.pdf.

8. Pinker, S. (2014). *The Sense of Style: The Thinking Person's Guide to Writing in the 21st Century*. Viking Adult.

9. Schimel, J. (2011). *Writing Science: How to Write Papers That Get Cited and Proposals That Get Funded*. Oxford University Press, New York.

10. Turabian, K. L., Booth, W. C., Colomb, G. G., Williams, J. M., and the University of Chicago Press Staff. (2013). *A Manual for Writers of Research Papers, Theses, and Dissertations*, Eighth Edition. University of Chicago Press (2013). DOI: 10.7208/chicago/9780226816395.001.0001.

2.2.7 SHARPEN YOUR SPEAKING SKILLS

Giving presentations on your campus, at leading centers of research for your field, or at businesses is one of the most effective ways of promoting your work, getting feedback, and building continuing relationships with colleagues at these centers. The best advice I can give is to practice, practice, practice. Every year, two weeks before the annual symposium for our lab, we hold three days of practice. Every one of the 25+ speakers goes through a supportive community discussion, whereby students and senior faculty deliver their 10–15-min talks, followed by 15 min of discussion and suggestions.

The University of Maryland (May 2017).

Just forcing speakers to prepare their slides early and to practice has an enormous impact in improving their talks. Beyond ensuring that each deck includes the basics—key people, sponsors, the webpage for their project, the date, and an email address—these sessions reinforce perhaps the most important piece of feedback for any presentation: closing with a clear takeaway slide. In other words: make it explicit what your key lessons are.

Most first-time presenters are understandably anxious, but practicing with a sympathetic audience helps get them past their fears. Some foreign students have difficulty in English, so I send them to our International Students Office, which puts them in touch with volunteer speech coaches.

You'll have to find your own style, but these guides can help.

1. Feamster, N. and Gray, A. (2013). Great Research: Presenting a Technical Talk: https://bit.ly/1kzCKG0.

2. Thimbleby, H. (2015). Pirate Talks: Audience, Remember, Route, Reflection (ARR). https://bit.ly/2P4ZtL6.

3. Carter, M. (2013). *Designing Science Presentations: A Visual Guide to Figures, Papers, Slides, Posters, and More*. Academic Press.

4. Anderson, C. (2013). How to Give a Killer Presentation. *Harvard Business Review*.

5. How to Give a Talk. Blog by David Stern. http://www.howtogiveatalk.com.

2.2.8 STRENGTHEN YOUR CAMPUS'S SOCIAL MEDIA PRESENCE

Social media, such as Twitter, Facebook, Instagram, Wikipedia, YouTube, and Vimeo, are powerful ways to reach targeted audiences. Some campuses have policies to coordinate their social media to standardize the style, content, and frequency of messages.

I worked with two prominent colleagues to develop Wikipedia pages on their work. They, in turn, worked with trusted students and colleagues to learn Wikipedia's style and include links that documented their awards and recognition. Then they set out to help other deserving colleagues create their own Wikipedia articles. The process helped my colleagues understand the difference between an academic biography and the expectations for a public encyclopedia. Only prominent faculty who have gained significant, external visibility are likely to have pages that will pass scrutiny by Wikipedia editors. One helpful feature of Wikipedia is that you can see daily pageviews, which show spikes when you get an award or your work is mentioned in the media.

Your campus research office probably has a Twitter or Facebook account, but you could work with them to increase the number of followers they have.

Videos are an increasingly powerful way to tell a story about research, but you'll need to practice your presentation carefully to get your message out. Then you'll need to publicize your video via your website and social media channels.

2.2.9 GENERATE TARGETED EMAIL LISTS

Traditional email lists are one of the most effective way to reach professionals. Developing targeted lists of people will enable you to announce your latest paper, invite regional professionals to a campus lecture, and/or offer advice to student project teams. Software like Google Groups and MailChimp make list maintenance painless by allowing individuals to sign up by themselves. You may want to personally manage special lists of close colleagues or potential corporate sponsors. Your lab, department, or college may generate larger lists, sometimes tied to automatic announcements of events entered in a central calendar.

2.2.10 CREATE AN AWARD FOR RESEARCH COMMUNICATIONS

You could arrange for awards for faculty and students who excel at telling about their research through media that reach the public, policymakers, or professionals. These media could consist of op-eds in major newspapers such as the *New York Times* or the *Washington Post*, education-oriented publications such as the *Chronicle of Higher Education* or InsideHigherEd.com. Local and national radio and television stations often welcome fresh news about research that addresses current issues. Other opportunities lie with prominent magazines such as *Scientific American*, *The Atlantic,* and *Psychology Today*.

At the University of Maryland, four faculty members each year earn the Research Communicator Impact Award (http://www.research.umd.edu/communicatoraward). While the early awards were just a certificate, here are a variety of ways your Vice President for Research might consider expanding this recognition:

- promote the work of awardees to local media;
- invite the awardees to speak at a workshop for other faculty about writing for general audiences;
- hold a luncheon among current and past awardees to discuss their work;
- offer awardees the chance to do featured talks about their work on campus;
- arrange for awardees to give talks off campus in the region; and
- produce a short video of the awardees explaining their work.

The University of Maryland at College Park (September 2017).

2.2.11 ORGANIZE RESEARCH FAIRS

Invite research partners from your campus, businesses, governments, and NGOs to establish a research fair to present their work to colleagues, campus administrators, and institutional leaders. It's best to create an event that focuses on one topic, like energy sustainability or healthcare. Be sure to invite campus public relations staff and local journalists, and advertise to regional organizations, nearby campuses, and students.

Research fair at the University of Maryland (May 2017).

2.2.12 PROMOTE TEAMWORK IN RESEARCH GROUPS

Teach the strategies for forming, participating in, and managing research teams. Teamwork often has a higher impact than single-author papers, because well-run teams do better, more ambitious work, and apply multiple research methods. Document your role in team projects, preferably in the papers you write.

1. Cooke, N. and Hilton, M. (Editors) (2015). *Enhancing the Effectiveness of Team Science*. National Academies Press, Washington, DC (2015). https://www.nap.edu/catalog/19007/enhancing-the-effectiveness-of-team-science.

2. Duhigg, C. (2016), What Google learned from its quest to build the perfect team. *New York Times*. https:/www.nytimes.com/2016/02/28/magazine/what-google-learned-from-its-quest-to-build-the-perfect-team.html.

3. Isaacson, W. (2014). *The Innovators: How a Group of Hackers, Geniuses, and Geeks Created the Digital Revolution*. Simon & Schuster, New York.

4. Johnson, S. (2010). *Where Good Ideas Come From: A Natural History of Innovation*. Riverhead Publishers.

5. Olson, J. S. and Olson, G. M. (2013). *Working Together Apart: Collaboration over the Internet*. Morgan & Claypool Publishers. DOI: 10.2200/S00542ED1V01Y201310HCI020.

6. Olson, G.M., Zimmerman, A., and Bos, N. (2008). *Scientific Collaboration on the Internet*. MIT Press, Cambridge, MA. DOI: 10.7551/mitpress/9780262151207.001.0001.

7. Shenk, J. (2014). *Powers of Two: Finding the Essence of Innovation in Creative Pairs*. Houghton Mifflin Harcourt.

Rensselaer Polytechnic Institute (February 2010).

2.3 BIGGER, LONG-TERM MISSIONS

If you've succeeded with the aforementioned efforts and projects, you may be ready to take on missions. These may produce resistance or intransigence from your colleagues, so you'll need to be resilient and persistent.

The University of Maryland (September 2016).

2.3.1 INCLUDE TEAM PROJECTS IN COURSES

Teach teamwork by having students work in teams on substantial projects to benefit someone outside the classroom. This way, the resulting projects can endure beyond the semester. There can be large benefits to students and instructors when teamwork is elevated to an essential role.

1. **MIT Teaching and Learning Lab:** http://tll.mit.edu/help/teaching-teamwork. Given the increased emphasis in universities on helping students acquire the skills they'll need to succeed professionally and personally, more instructors are experimenting with student teams. But it's

not enough to put students in groups and ask them to work together—students need to be taught how to function in this kind of situation.

2. **Teaching Teamwork (The Oxford University Press blog):** https://blog.oup.com/2016/07/teaching-teamwork/. Teamwork projects may require fresh thinking by faculty members, but it may be easier to supervise and grade 10 teams of four students, than to mentor and grade 40 individuals. Moreover, well-designed teamwork projects could lead to published papers or start-up companies in which faculty are included as co-authors or advisers.

3. **TEACH Teamwork (The American Psychological Association):** http://www.apa.org/education/k12/teach-teamwork.aspx. An evidence-based, self-guided program on how to work in teams.

4. **Teaching Teamwork (The University of Washington):** https://www.youtube.com/watch?v=eY2fG_Fbm2M. University students are often asked to work in groups, yet few are taught how to actually work this way. Unsurprisingly, they often struggle. University of Washington faculty Randy Beam (Communication) and Erin Hill (Physics) describe how they integrate instruction on teamwork into their regular courses to help students succeed in both short- and long-term class projects.

Harvard University (October 2014).

2.3.2 TEACH THE METHODS OF DESIGN THINKING

Design thinking is a new process that's different from engineering but could have a big impact on many science-oriented research projects. Students can learn about it on their own, but it's better if faculty create new courses. Meanwhile, academic leaders can tune these courses to the needs of different disciplines. Design degrees or minors will also raise your campus's design thinking competence. Among the definitions of "design thinking," I favor Tim Brown's (IDEO) short definition:

> *"Design thinking is a human-centered approach to innovation that draws from the designer's toolkit to integrate the needs of people, the possibilities of technology, and the requirements for business success."*

While there are many versions of design thinking processes, I like Aaron Marcus's sequence:

- plan project;

- analyze needs;
- gather requirements;
- design initial solution;
- evaluate design solutions (iterate with initial and revised design steps);
- design revised solution;
- evaluate design concepts (iterative);
- deploy product/service;
- evaluate product/service (iterative);
- determine future requirements/enhancements;
- maintain and improve processes; and
- assess project.

Here are some further sources on design thinking:

- https://en.wikipedia.org/wiki/Design_thinking;
- http://dschool.stanford.edu/dgift/;
- http://designthinking.ideo.com;
- http://www.designthinkingforeducators.com; and
- Dam, R. and Siang, T. (2018). What is Design Thinking and Why Is It So Popular? https://www.interaction-design.org/literature/article/what-is-design-thinking-and-why-is-it-so-popular.

The University of California at Berkeley (October 2016).

2.3.3 RAISE TWIN-WIN ISSUES AT CONFERENCES

Most students and faculty are used to talking about their results at conferences that focus on specific research topics, like cybersecurity or diabetes treatment. These events serve an important purpose, but you might consider conferences devoted to academic topics like X and Y. These conferences are organized by professional societies that serve the academic community, and would likely welcome a talk about your approach to Twin-Win research. Here's a list of organizations you can consult:

1. **Association of Public and Land-grant Universities:** http://www.aplu.org/hibar;
2. **Business-Higher Education Forum:** http://www.bhef.com/about;
3. **Coalition for Networked Information:** https://www.cni.org;
4. **EDUCAUSE:** https://www.educause.edu;
5. **Government-University-Industry Research Roundtable:** http://sites.nationalacademies.org/PGA/guirr/PGA_081653;

6. **National Alliance for Broader Impacts:** https://broaderimpacts.net/about/;
7. **National Organization of Research Development Professionals:** http://www.nordp.org/conferences; and
8. **University-Industry Demonstration Partnership:** https://www.uidp.org.

The University of British Columbia (September 2016).

2.3.4 COLLECT EVIDENCE OF TWIN-WIN PAYOFF

Convincing students, faculty, and academic leaders of the value of Twin-Win thinking is usually successful when you have a combination of powerful success stories and objective data. The stories can showcase breakthrough discoveries or positive societal impacts. The data can include bibliometric information such as the number of papers published, paper downloads, and citations. Other data include the number of patents, start-up companies, licenses, presentations, and mentions in social or public media. In some cases, there are other metrics such as software download counts or use of your data by others. Evidence of societal benefits are much harder to track, since there are usually many factors that lead to a broad implementation of a research result.

A useful model could be the United Kingdom's Research Excellence Framework (http://www.ref.ac.uk/about/), which produced a national evaluation in 2014 and will do so again in 2021. One remarkable aspect of the 2014 evaluation (http://www.hefce.ac.uk/rsrch/REFimpact/) was the database of 6,975 case studies that showed how research projects bring societal benefits.

The University of Minnesota (February 2011).

2.3.5 LEARN ABOUT ALTERNATE WAYS TO ASSESS THE IMPACT OF YOUR WORK

The traditional way to claim research success is by counting the number of papers you published, especially in high-quality journals. Improvements in online databases that include 70 million research papers have made it possible to count the number of citations for each paper, and often the numbers of downloads.

Each of these metrics can be "gamed," or manipulated to inflate impact, and there are vast differences in numbers of publications and number of authors for papers. For example, some physics, astronomy, and biology papers have more than a thousand authors, who then each cite their joint paper, thereby producing a high number of citations and downloads. Some fields are large with many papers and many researchers, whereas other fields are small with few papers and authors. This means that evaluations have to take into account the size of each field.

Another problem is that some work has an outsize impact in business or government but produces very few citations. A strong movement has arisen to promote qualitative approaches, such as the Becker Model (https://becker.wustl.edu/impact-assessment/model).

Here are a few alternate metrics.

1. Galligan, F. and Dyas-Correia, S. (2013). Altmetrics: Rethinking the way we measure. *Serials Review* 39, 56–61. DOI: 10.1080/00987913.2013.10765486.
2. http://www.altmetric.com/top100/2014/.
3. http://www.altmetrics.org.
4. Penfield, T., Baker, M. J., Scoble, R., and Wykes, M. C. (2014). Assessment, evaluations, and definitions of research impact: A review. *Research Evaluation* 23, 21–32. DOI: 10.1093/reseval/rvt021.

2.3.6 CHANGE HIRING, TENURE, AND PROMOTION POLICIES TO ENCOURAGE AND RECOGNIZE TEAMWORK

Since academic hiring, tenure, and promotion policies focus on individual accomplishment, adjustments are necessary to address the growing number of situations in which an individual is part of a research team. For example, the University of Southern California has developed guidelines to allow candidates to describe their collaborative scholarship that contributes to larger projects. Increasingly, journals and conferences require descriptions of each co-author's role in preparing the paper.

- University of Southern California, Guidelines for Assigning Authorship and for Attributing Contributions to Research Products and Creative Works (September 16, 2011). https://bit.ly/2oe24XM.

• Moher, D., Naudet, F., Cristea, I. A., Miedema, F., Ioannidis, J. P. A. , and Goodman, S. N. (2018). Assessing scientists for hiring, promotion, and tenure. *PLOS Biology* 16(3): e2004089. DOI: 10.1371/journal.pbio.2004089.

Penn State University (2009).

2.3.7 CHANGE HIRING, TENURE, AND PROMOTION POLICIES TO ENCOURAGE AND RECOGNIZE WORKING WITH BUSINESS, GOVERNMENT, AND NGO PARTNERS

In assessing research contributions, more than 45 North American universities now count patents as equivalent to published papers. The seven-year-old National Academy of Inventors honors academics who make inventions (http://www.academyofinventors.org/about.asp) that "enhance the visibility of academic technology and innovation… and translate the inventions of its members to benefit society."

1. **Texas A&M Will Allow Consideration of Faculty Members' Patents in Tenure Process:** http://www.chronicle.com/article/Texas-A-M-Will-Allow/118262/.

2. **Association for Public and Land-grant Universities Task Force on Tenure, Promotion, and Technology Transfer Survey Results and Next Steps:** https://bit.ly/2MS612l.

3. **U of Maryland to Count Patents and Commercialization in Tenure Reviews:** http://www.chronicle.com/article/U-of-Maryland-to-Count/132261.

4. Stevens, A. J., Johnson, G. A., and Sanberg, P. R. (2011). The role of patents and commercialization in the promotion and tenure process. *Technology & Innovation* 13, 3, 241-248. DOI: https://doi.org/10.3727/194982411X13189742259479.

5. Sanberg, P.R., Gharib, M., Harker, O. T., Kaler, E. W., Marchase, R. B., Sands, T. D., Arshadi, N., and Sarkar, S. (May 2014). Changing the academic culture: Valuing patents and commercialization toward tenure and career advancement. *PNAS Perspectives*. http://www.pnas.org/content/111/18/6542.

2.3.8 INCORPORATE NEW GOALS AND MEASURES IN STRATEGIC PLANS AND VISION STATEMENTS

Clarify the role of research in contributing to civic, business, and global priorities. Encourage researchers to take on projects that make contributions to each campus, local communities, cities, counties, and states. Increase the prominence of campus Offices of Technology Commercialization and entrepreneurial programs such as the NSF I-Corps. Develop best practices and hold conferences to share experiences.

Here are some sample strategic plans:

1. **City College of New York:** https://www.ccny.cuny.edu/it/strategic_planning;

2. **Cornell University:** https://www.cornell.edu/strategicplan/;

3. **Honolulu Community College:** https://www.honolulu.hawaii.edu/strategicplan;

4. **Northwestern University:** http://www.northwestern.edu/strategic-plan/;

5. **University of California at San Diego:** http://plan.ucsd.edu;

6. **University of Illinois:** http://strategicplan.illinois.edu;
7. **University of Maryland:** https://www.umd.edu/strategic-plan; and
8. **University of North Carolina:** https://www.northcarolina.edu/strategic-planning.

2.3.9 WORK WITH YOUR CAMPUS'S OFFICE OF TECHNOLOGY COMMERCIALIZATION

Most campuses have an office that licenses campus intellectual property such as books, software, music, and survey instruments. The traditional model was to focus on these transactions, but there's a growing awareness that these offices can also develop long-term relationships with the campus. The benefits of these relationships include research collaborations, student internships, faculty sabbaticals, and faculty hiring. Consider relationships with businesses. Such partnerships can bring many benefits to campus researchers, such as identifying problems, connecting with business researchers, using specialized equipment or data, and providing testbeds for solutions.

Stanford University's Office of Technology Licensing (OTL, http://otl.stanford.edu) has a traditional, narrow definition of its goals: "The Office of Technology Licensing evaluates, markets and licenses technology owned by Stanford. OTL's mission is to encourage effective technology transfer for the public benefit as well as generating royalty income for Stanford to benefit research and education."

The University of Maryland's Office of Technology Commercialization (OTC, http://otc.umd.edu/industry) says that it helps "fuel research and entrepreneurial initiatives through inter and intra-university collaborations. OTC is proud to act as a liaison between university members and the business community to further these projects."

Stanford University (May 2008).

2.3.10 EXPAND COLLABORATION WITH BUSINESS

To smooth the way, develop standard agreements that resolve issues of intellectual-property sharing and which support prompt publication of novel research results. Encourage local and regional funding that supports researchers who work with civic partners in local communities, cities, counties, and states. Give awards for the best off-campus partnerships, and tell success stories. Build stronger partnerships with off-campus civic, business, and global partners who bring genuine problems.

Some businesspeople appreciate the unique nature of academic research and they come prepared to work with faculty and students in productive ways that lead to papers, student projects, and long-term research. Other relationships may take time to develop into productive partnerships. Spending days or weeks learning about what businesses need may be satisfying to faculty and students. Small projects can mature into bigger partnerships that produce the Twin-Win results, which have high payoffs for everyone. Partnerships with major companies have led to substantial funding for our research, opportunities to consult, and even new products. Connections with small local companies have led to a chance to help them grow, jobs for students, and sponsorship for our annual symposium.

1. The Computing Research Association, Computing Community Consortium. (2016). *The Future of Computing Research: Industry-Academic Collaborations*. http://cra.org/ccc/wp-content/uploads/sites/2/2016/06/15125-CCC-Industry-Whitepaper-v4-1.pdf.

2. Etzkowitz, H. (2008). *Triple Helix: University, Industry Government Innovation in Action*. Routledge Publishers, New York. DOI: 10.4324/9780203929605.

3. 10 Case Studies of High-Value, High-Return University-Industry Collaborations, UIDP (2014). https://www.uidp.org/wp-content/uploads/documents/Case-Studies-pre-20141.pdf.

4. The Business-Higher Education Forum. http://www.bhef.com.

5. The American Association of University Professionals. (2014). 56 Principles to Guide Academic-Industry Relationshps. https://www.aaup.org/sites/default/files/files/Principles-summary.pdf.

The University of North Carolina at Charlotte (March 2011).

2.3.11 EXPAND COLLABORATION WITH CIVIC PARTNERS

Reaching out to city, state, regional, and national government agencies could bring great opportunities for Twin-Win research. Agricultural and technology extensions are the front door of local government.

1. **MetroLab Network:** https://metrolabnetwork.org. MetroLab includes 41 cities, 4 counties, and 55 universities, organized in more than 35 regional city-university partnerships. Partners focus on research, development, and deployment projects that offer technological and analytically based solutions to challenges facing urban areas, including inequality in income, health, mobility, security and opportunity; aging infrastructure; and environmental sustainability and resiliency.

2. **10 Principles for Successful City-University Partnerships:** https://metrolabnetwork.org/primer/. Embrace the idea of the city as a "living lab" and the university as an R&D resource where faculty and students can work on policies and technologies that enhance quality of life and advance the understanding of cities and urban science.

3. **New York University's (NYU) Center for Urban Science and Progress (CUSP):** http://cusp.nyu.edu. CUSP is a university-wide center whose research and education programs focus on urban informatics. Using New York City as its lab, and building from its home in NYU's Tandon School of Engineering, CUSP integrates and applies NYU strengths in the natural, data, and social sciences to understand and improve cities throughout the world.

4. **Massachusetts Institute of Technology's Center for Civic Media:** https://civic.mit.edu. Working at the intersection of participatory media and civic engagement, the Center for Civic Media seeks to design, create, deploy, and assess tools and processes that support and foster civic participation and the flow of information between and within communities.

5. **University of California's Center for Information Technology Research in the Interest of Society (CITRIS) and the Banatao Institute:** http://citris-uc.org. CITRIS and the Banatao Institute create IT solutions for society's most pressing challenges.

2.3.12 SPREAD TWIN-WIN IDEAS IN PROFESSIONAL SOCIETIES

For academic leaders who are involved with professional societies, journals, and conferences, you can win them over to Twin-Win ideas by helping them revise their descriptions of goals, coverage of journals, and topics for conferences. Some conferences have a government and industry track that seeks research that's demonstrated high impact on meaningful problems. These tracks need to have separate review panels that are given explicit instructions to respect these novel kinds of papers. As more societies, journals, and conferences shift their emphasis to embrace Twin-Win ideas, authors will find it easier to gain acceptance for their work. Eventually, traditional papers that lack linkage to meaningful problems will become less acceptable.

2.3.13 ENCOURAGE RESEARCH PROGRAMS THAT COMBINE THEORY AND PRACTICE

Research-funding agencies are increasingly sympathetic to HIBAR and Twin-Win ideas. For example, the National Science Foundation's (NSF) Algorithms in the Field (AitF, http://www.nsf.gov/pubs/2015/nsf15515/nsf15515.htm) program encourages inclusion of a theory person and domain expert in research proposals. Similarly, NSF's Big Data Regional Innovation Hubs (http://www.nsf.gov/pubs/2015/nsf15562/nsf15562.htm) describes a happy blend of research strategies that resembles the successful ideas of agricultural and technology extension services.

CHAPTER 3
Conclusion

The paths I've described in this guidebook are meant as a starting set of ideas to change your research, transform your campus, and, ideally, produce wider reforms. Some campuses have seed grant programs to give faculty and students opportunities to work on new ideas. Another way of producing changes is by way of "cluster hires," which bring several new faculty working on related topics to different colleges, so as to trigger interdisciplinary, problem-driven research. Some of you may also know journalists who you could engage with to write about this growing and transformational movement.

For those of you excited by these possibilities, please send me your suggestions for other paths that you use on your campus. I'm available at ben@cs.umd.edu. Please write up your efforts and publish them on the web for others to learn from. Together, we can produce meaningful and sustained changes that steer research to ever-greater benefits to society.

Stony Brook University (May 2015).

NOTES

Where can you have the most impact in the shortest time?

Choose from the small, medium, and large efforts.

Who will be your partners in creative change?

What peers, superiors, and staff can you rely on?

What is your schedule for moving forward?

Do something small soon, but get started.

How will you assess your progress?

Can you count the number of people you talk with or send emails to?

How might your effort fail?

Anticipating dangers will strengthen your efforts.

As your effort moves forward, can you step back and let others lead?

As your small effort grows, be sure to share leadership and success.

Author Biography

Ben Shneiderman (http://www.cs.umd.edu/~ben) is a Distinguished University Professor in the Department of Computer Science, Founding Director (1983–2000) of the Human-Computer Interaction Laboratory (http://hcil.umd.edu), and a Member of the UM Institute for Advanced Computer Studies (UMIACS) at the University of Maryland. He is a Fellow of the AAAS, ACM, IEEE, and NAI, and a Member of the National Academy of Engineering, in recognition of his pioneering contributions to human-computer interaction and information visualization. His innovative contributions include the web's highlighted link that makes it easy for billions of users to get the information they want and the tiny touchscreen keyboard on mobile devices used around the world. His theories, research methods, and software tools have become popular topics in computer science, while revolutionizing the ways people use technology to improve their lives. He has received six honorary doctorates.

Shneiderman's recent books are *Designing the User Interface: Strategies for Effective Human-Computer Interaction* (6th ed., 2016) and *The New ABCs of Research: Achieving Breakthrough Collaborations* (2016).

Printed in the United States
by Baker & Taylor Publisher Services